Greetings, Little Stargazers!

Get ready for a fun journey to the stars and beyond! In this coloring book, we'll dive into the amazing world of the 2024 Total Solar Eclipse.

As you color these pages and create some cool artwork, you'll also become a true eclipse expert and meet some cute little friends along the way! You'll learn about the Sun, Moon, Earth, and how they all come together to create this super special event. Let the cosmic coloring begin!

. . . . ✦ . ✦ . ✦ . . . .

Hey, Cosmic Parents! Follow our Facebook page 'StoriLand' and message us to get 5 free printable coloring pages! Plus, you'll also be able to stay up to date on future freebies and new books! Your gateway to galactic fun is at: https://www.facebook.com/StoriLand

. . . . ✦ . ✦ . ✦ . . . .

Copyright © 2024 by StoriLand

All rights reserved. No part of this book may be reproduced in any manner. This book is for personal use only and commercial use is strictly prohibited.

StoriLand

Total Solar Eclipse Fun Facts

☆ **A Total Solar Eclipse** happens when the Moon moves between the Earth and the Sun. It blocks the Sun's light for a little while, making it dark during the day. This can only happen when the Moon, Sun, and Earth are lined up perfectly.

☆ **Full Moon:** The Moon has to be in its "full" phase, looking like a complete circle, to create a total solar eclipse.

☆ **Minutes:** During a total solar eclipse, the Moon moves really fast across the sky. It only takes a few minutes for the whole show to happen.

☆ **Perfect Fit:** The Moon is about 400 times smaller than the Sun, but it's also about 400 times closer to Earth. This cosmic coincidence makes the Moon look just the right size to cover the Sun during a total solar eclipse.

☆ **Orbital Moves:** The Moon's distance from Earth isn't always the same because its path in space can be a little wavy. Sometimes, it's closer, and other times, it's farther away. But when it lines up just right with the Sun, we get to see a total solar eclipse!

It's important to wear safety glasses during an eclipse to keep our eyes safe from the bright Sun.

5 STAGES OF A TOTAL SOLAR ECLIPSE

1

First contact marks the beginning of a partial eclipse as the Moon slowly moves in front of the Sun. It may look like the Moon is taking a little bite out of the Sun.

2

Second contact is when the Moon begins to cover the whole Sun. Some sunlight will still peek through from around the Moon.

3

Totality is the stage when the Moon covers the Sun completely. This is the most exciting part because the sky goes dark, and even the birds and animals become quiet.

4

Third contact is when the Moon begins to move away from the Sun, allowing it to become visible again as sunlight starts peeking through. The total eclipse is coming to an end.

5

Fourth contact is when the Moon moves even farther away from the Sun, no longer covering it, marking the end of the eclipse.

Changing Temperatures: During a total eclipse, the temperature can drop, and it might feel cooler, almost like a sudden breeze. This happens because the Moon blocks some of the sunlight, making it a bit colder for a few minutes.

Totality Path: The path of totality is the area on Earth where people can see a total solar eclipse. It's usually a narrow path, so not everyone around the world gets to experience totality. However, many people outside this path will still be able to see a partial eclipse. The curved line over Earth shows the path of totality for the 2024 eclipse, which includes areas of Canada, the United States, and Mexico.

Total solar eclipses are super special because they don't happen very often, maybe only once every year and a half to two years.

Scientists and astronomers get really excited during a total solar eclipse because they can learn cool things about the Sun and space. Some of the things they study include:

★ **Solar Corona:** Scientists study the Sun's outer atmosphere, called the corona. It's like a shiny halo around the Sun. Eclipses let them see it better because it's usually too bright to look at.

★ **Sun's Surface:** They also examine the Sun's surface, called the photosphere, during an eclipse. This helps them learn more about sunspots and the Sun's surface temperature.

★ **Sun's Magnetic Field:** Scientists use special tools during an eclipse to check the Sun's magnetic field, like a giant invisible magnet. This helps them understand how the Sun's energy affects Earth.

Fun Facts About The Sun

★ **Giant Star:** The Sun is a giant star! But since it's the closest star to us, we call it the Sun.

★ **Big Size:** The Sun is so big that more than a million Earths could fit inside it.

★ **Light Travel:** It takes about 8 minutes for sunlight to reach us from the Sun.

★ **Solar System's Boss:** The Sun is the boss of our solar system. It keeps all the planets, including Earth, in orbit with its gravity.

★ **Life Giver:** Without the Sun, there would be no life on Earth. It gives us warmth and energy for plants to grow and for us to live.

★ **Millions of Years Old**: The Sun is really, really old. It's been shining for about 4.6 billion years.

Fun Facts About The Earth

☆ **Blue Planet:** Earth is often called the 'Blue Planet' because, when seen from space, it looks mostly blue. That's because more than 70% of Earth is covered in water, not only in oceans but also in rivers, lakes, and even underground.

☆ **Spinning Globe:** It takes Earth about 24 hours to complete one full spin. This spinning motion is what gives us day and night. When one side of Earth faces the Sun, it's daytime, and when it turns away, it becomes nighttime!

☆ **Earth's Journey:** Earth takes about 365 days to travel all the way around the Sun. That's why we have 365 days in a year! It's like a grand cosmic journey, and each day is a step in our amazing trip around our star, the Sun!

Fun Facts About The Moon

✦ **Moon Phases:** The Moon looks different every night! Sometimes it's round and full, and other times it's a crescent, shaped like a banana. This happens because we see different parts of it as it goes around the Earth.

✦ **Moon's Craters:** The Moon's surface is covered in big holes called craters. They're like cosmic potholes! They were made by things like meteors crashing into the Moon.

✦ **Glow in the Dark:** The Moon doesn't make its own light, but it shines because it reflects the Sun's light.

Fun Facts About Meteors

☆ **Shooting Stars:** Meteors are sometimes called "shooting stars," but they're not really stars at all. They're chunks of rock or metal that zoom through space. When they enter Earth's atmosphere, they move super-fast and the friction of the air causes them to heat up and catch on fire, creating those bright streaks of light.

☆ **Meteor Showers:** Sometimes, lots of meteors appear in the sky at once, and we call this a "meteor shower." It's like a cosmic fireworks show.

☆ **Meteoroids, Meteors, and Meteorites:** Before they hit Earth, these space rocks have different names. They are called "meteoroids" when they're in space, "meteors" when they streak through the sky, and "meteorites" if they land on Earth.

☆ **Speedy Travelers:** Meteors can travel really, really fast. Some can zoom through space at over 160,000 miles per hour (257,500 kilometers per hour).

Meteors Glow in Different Colors

Meteors are made of various minerals and chemicals, and when they react with the Earth's atmosphere, it causes them to appear in the beautiful colors we see as they streak across the sky.

www.ingramcontent.com/pod-product-compliance
Lightning Source LLC
Chambersburg PA
CBHW080959290526
45795CB00009B/3009